Cristina Ivaldi

Monster in the box

商務印書館

Contents

Originally published by Black Cat Publishing under the title: *Monster in the box*
© 1999 Black Cat Publishing
An imprint of Cideb Editrice, Genoa, Canterbury

The copyright of this Chinese edition is owned by
The Commercial Press (H.K.) Ltd.

Name of Book: Monster in the box
Author: Cristina Ivaldi
Editors: Claudia Fiocco, Monika Marszewska, Elvira Poggi Repetto
Design and Art Direction: Nadia Maestri
Illustrations: Alfredo Belli
Layout: Sara Blasigh

系 列 名： Quality English Learning for Kids · I
書　　名： Monster in the box
責任編輯： 傅　伊
出　　版： 商務印書館（香港）有限公司
　　　　　香港筲箕灣耀興道 3 號東滙廣場 8 樓
　　　　　http://www.commercialpress.com.hk
印　　刷： 美雅印刷製本有限公司
　　　　　九龍觀塘榮業街 6 號海濱工業大廈 4 樓 A
版　　次： 2004 年 9 月第 1 版第 1 次印刷
　　　　　© 2004 商務印書館（香港）有限公司
　　　　　ISBN 962 07 1720 1
　　　　　Printed in Hong Kong

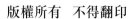

出版說明

　　學英語當然要學優質的，有品質才能讓人有信心。我們一直積極提倡學習優質英語的理念，並且為學習者提供過多元化的優質英語材料，像《Black Cat 優質英語階梯閱讀》就十分成功，至今已出版近 60 本。鑑於良好的英語能力最好從小培養，我們於是出版這一套適合五至八歲兒童的優質英語閱讀讀本 "Quality English Learning for Kids・I"。

　　培養兒童對於英語的興趣，須從趣味和簡易兩方面入手。圖文並茂，聲文結合這兩大特點對學習英語甚有幫助。"Quality English Learning for Kids・I" 承續本館出版優質英語書的理念，全書彩圖精美，附 CD 朗讀內容及聆聽練習，形式多元化，有出版故事讀本（story books）、圖畫讀本（picture readers）、戲劇讀本（drama readers）及互動讀物（interactive readers）四大類，提供不同學習功能。故事讀本和圖畫讀本可供兒童看圖講故事；戲劇讀本完全用對白編寫，培養脫口而出講英語的習慣，適合家庭裏作簡單的角色扮演，或者小學生在課堂上作簡單的演出。

　　針對兒童學習英語的需要，本系列提示家長為兒童設定學習目標，並且說明如何達標，另備生詞表和語法知識點，讓兒童在家長協助下掌握生詞用法，認識簡單的句子結構和了解語法要點。

　　"Quality English Learning for Kids・I" 吸引兒童對閱讀產生興趣，逐步引導他們參與愉快的閱讀旅程。在這個旅程中，家長是重要的導航者，透過對兒童的悉心鼓勵，循循善誘，進一步加強親子關係。

商務印書館
編輯部

使用説明

① 如何使用本書?

本書為圖畫讀本 (picture reader),適合課堂使用或親子共讀。

每頁均圖文並茂。正文包括對話和敘述文字,讓小孩子熟悉並掌握第一、二、三身句法。老師或家長可讓小孩子用第一身敘述句法作自我介紹,用第二身疑問句法向自己提問,用第三身敘述句法介紹親人、朋友或描述常見的動物。老師或家長還可與小孩子進行簡單對話,練習使用第一、二、三身句法。

除正文外,還設有看圖練習題 (辨識、連線、填色、字謎等),培養小孩子的觀察和判斷能力。老師或家長可指出實物,讓小孩子説出它們的顏色,或讓小孩子説出自己身體各部位的名稱和常見動物的各身體部位的名稱。

本書配有CD,小孩子可邊聽邊讀,提高英語聽説能力。

② 本書的學習目標是甚麼?

老師或家長可為孩子定出以下學習目標。

使用本書後,孩子學會:

(a) 説出身體各部位的名稱 (say parts of the body);
(b) 識數 (count numbers from 1 to 10);
(c) 辨別故事中不同角色 (identify different characters);
(d) 聽從CD的指示,玩識圖 (Labelling & Who are they)、填色 (Colour it)、連線 (Listen and join) 等遊戲。

③ 本書有哪些重點生詞和語法學習項?

(a) 重點生詞:本書的重點生詞包括四大類,即顏色 (colours)、數字 (numbers)、身體部位 (parts of the body) 以及描述人和事物情狀的形容詞 (adjectives)。另附英漢對照生詞表,增強對生詞的記憶。

(b) 語法學習項:
第一身句法 (the first person) (例如頁3,"I am great!")
第二身句法 (the second person) (例如頁4,"How old are you?")
第三身句法 (the third person) (例如頁3,"Bobo is a little boy.")
wh-疑問句 (wh-questions),主要由 "疑問詞+動詞+主語" (wh-word + verb + subject) 組成。(例如頁8,"What is it?" 和頁12,"What colour is it?")
祈使句 (imperatives),用於表示命令、請求等,主要由 "動詞+賓語" (verb + object) 組成。(例如頁7,"Go away, little..." 和頁16,"Open the box, Bobo!")

Bobo is a little boy.
He has got short black hair
and he wears glasses.
He is very nice.

1. HELLO!
MY NAME'S BOBO.

2. THESE ARE
MY GLASSES.

3. WOW!
I'M GREAT!
I'M HANDSOME
LIKE BRUCE
WILLIS.

Bobo is eight,
but he is little
for his age.
He is very short
and very thin.

4. I'M
EIGHT TOO.

5. EIGHT?
YOU?

Some boys at school are bullies,
but Bobo is not scared...

...because he has got his box.
It is a big white box.
It is under his arm.

This box is a mystery.
What is in the box?
Bobo says it is his friend.
What friend?

It is big.
It has got eight legs,
but it is not a spider.

It is pink
and blue
and yellow.

1. WHAT COLOUR
IS IT?

2. IT'S PINK
AND BLUE
AND YELLOW.

It has got a long tail
and a trunk.

A tail and a trunk?
But what is it?
A monster, of course!
His name is Jeremy.

Big Macho wants to open the box.
He wants to look inside the box.
But... it is dangerous!

1. OPEN THE BOX,
BOBO!

Bobo falls down.
The box falls.

Quality
English
Learning
For Kids

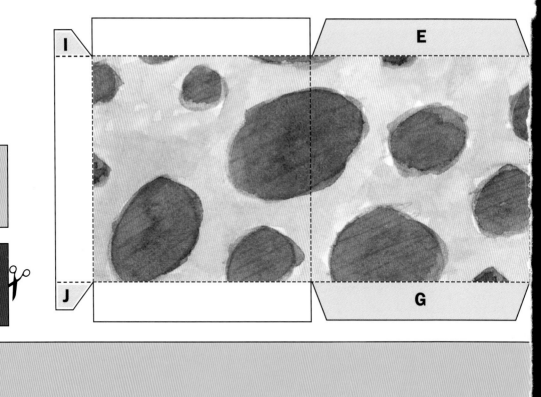

I

E

J

G

D

D

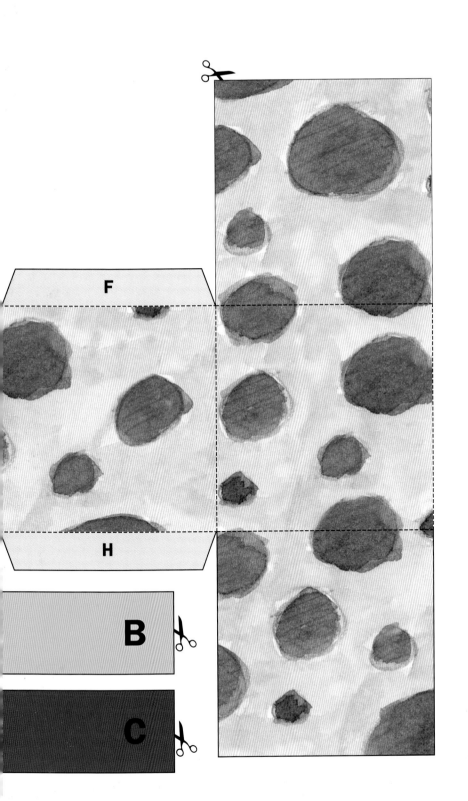

F

H

B

C

Jack in the box

1. Cut ✂ the 4 strips ▭ along the lines ———.
2. Glue [GLUE] B to B ▭ and C to C ▭.
3. Glue [GLUE] A to A. ▭
4. Fold the 2 strips together.
5. Glue [GLUE] D to D.
6. Pull ◢◣◢◣ the spring. 〰
7. Push ✋ the spring back.
8. Cut ✂ the box piece ▭ along the lines ———.
9. Fold 🗄 along the dashed lines --------.

B	

C	

A	

A	

10. Glue  E, F, G, H, I, J and make the box.

11. Cut Jack along the lines ——— .

12. Glue the spring to Jack (D to D) and to the bottom of the box.

13. Close the box.

14. Open the box.

D

In the box there is
a pink, blue and yellow toy monster
with eight legs,
a tail and a trunk.

OH, I SEE.
A *TOY* MONSTER.

No! It is not a toy!
It is a real monster!
Bobo is not scared,
but the boys
are very scared!

Bobo thanks Jeremy.

Jeremy is a real monster
and a real friend too.

THE END

23

Labelling

Listen and point to the parts of the body. Then use the words to label the parts of the boy.

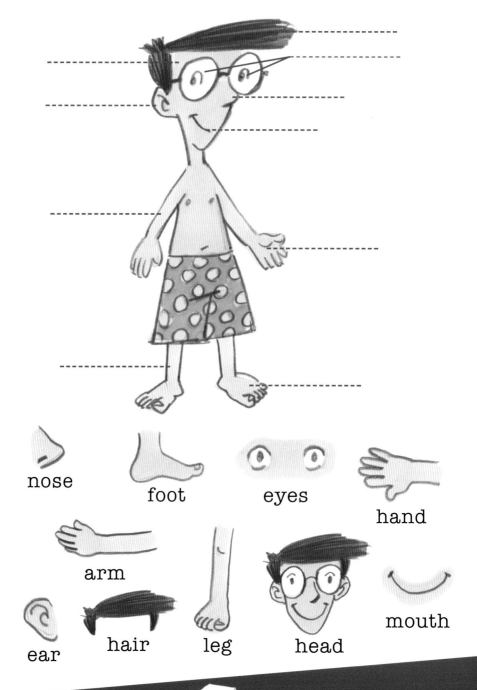

nose   foot   eyes   hand

arm   head   mouth

ear   hair   leg   head

4 Listen and point to the parts of the body. Then use the words to label the parts of the monster.

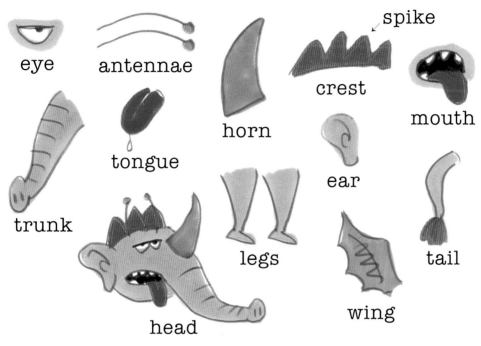

spike

eye   antennae

crest

mouth

horn

tongue   ear

trunk

legs   tail

wing

head

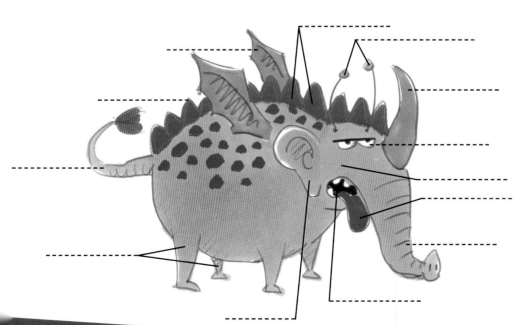

Crossword

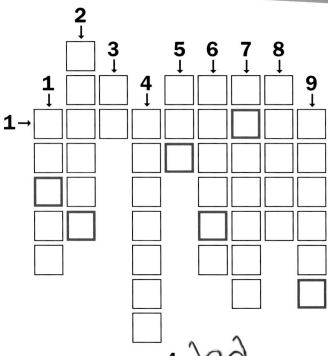

Across →

1 Don't open the box! It's .... .

Down ↓

1

2 Bobo isn't .... of Jeremy.

3 There's a monster .... the box.

4

5

6 Jeremy is a good .... for Bobo.

7

8 Big Macho is a real .... .

9

Anagram: Find the mysterious monster in the red boxes.

D __ __ __ __ __

Colour it!

 5 Listen to the instructions to colour the numbers.

 white

 pink

 purple

 brown

 green

 black

 blue

 orange

 red

 yellow

Who are they?

6 These monsters are Jeremy's friends. They are called John, Jeff, Jack, Jenny, Jessica, Jill and James. Listen and identify the monsters. Write their names in the boxes.

J......................

J......................

J......................

J......................

J...................

J......................

J......................

Look for the monsters!

7

Can you see Bobo? This is his classroom.
It is a funny classroom with a lot of monsters!
Where's Jeremy?

Now look for:

a monster with two heads
a monster with three arms
a fat monster
a tall monster

a monster with ten legs
a purple monster
a girl with blue hair
a boy with two pairs of glasses
a boy with red eyes

Listen and join!

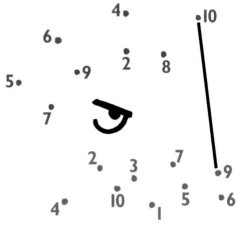

🎧 **8** Listen to the recording and join the dots.

4•   •10

6•

5•   •9   •2   •8

7

2•   3   •7   •9

4•   10   •1   5   •6

6•—•1   •3

2   •5   8•—•5

1•   •4   7   •9

6•   •4

3•   •3

9•

•1

8   6

7•   •2

•10   •4

•5   •1   •8

3•   2•—•10

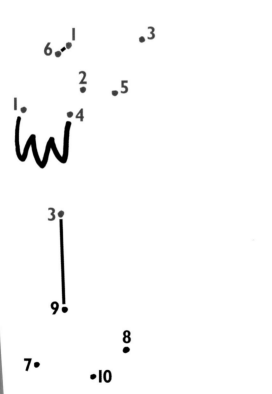

32

Labelling

pages 24 and 25

Listen and point to the parts of the body. Then use the words to label the parts of the boy.

nose; ear; arm; hair; foot; leg; eyes; head; hand; mouth.

Listen and point to the parts of the body. Then use the words to label the parts of the monster.

eye; trunk; antennae; tongue; head; horn; legs; crest; spike; ear; wing; mouth; tail.

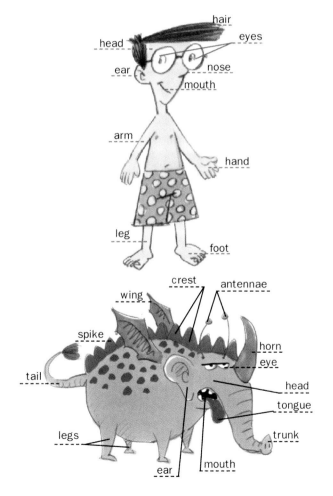

Crossword

page 26

Across
1 Don't open the box! It's **dangerous**.
Down
1 **Dwarf**
2 Bobo isn't **scared** of Jeremy.
3 There's a monster **in** the box.
4 **Glasses**
5 **Leg**
6 Jeremy is a good **friend** for Bobo.
7 **Monster**
8 Big Macho is a real **bully**.
9 **Spider**

Anagram: Find the mysterious monster in the red boxes.
DRAGON

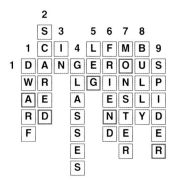

5 Colour it!

page 27

Listen to the instructions to colour the numbers.
Colour number one white.
Colour number two yellow.
Colour number three orange.
Colour number four red.
Colour number five blue.
Colour number six brown.
Colour number seven black.
Colour number eight purple.
Colour number nine pink.
Colour number ten green.

6 Who are they?

pages 28 and 29

These monsters are Jeremy's friends. They are called John, Jeff, Jack, Jenny, Jessica, Jill and James.

John is red and blue. He's got two ears, two eyes and four legs. He's got two orange horns on his head. He hasn't got a tail.

James is red. He's big and he's got two trumpet ears, two eyes, four legs and a lot of spikes. They're green, blue, purple, red, orange and yellow.

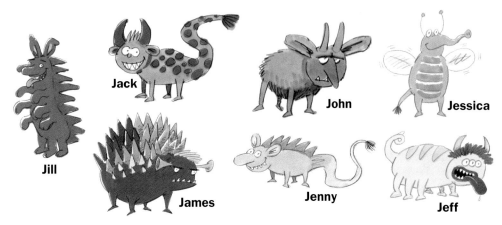

Jack

John

Jessica

Jill

James

Jenny

Jeff

Jill is purple. She's got nine legs, two eyes, two noses and two long ears. She's got a crest on her back.

Jessica is green and blue. She's got two legs. She's got two antennae and two eyes but she hasn't got any ears. She's got two wings, a tail and a trunk. She looks like a bug.

Jeff is yellow. He's got six legs, two horns, three eyes and a big mouth with a long tongue. He's got purple hair.

Jenny is green. She's got four legs, two ears, two eyes and a big nose. She's got a long yellow tail and a crest on her back.

Jack is pink and purple. He's got two ears, two eyes and four legs. He's got a long tail. He's got two horns on his head.

⟨7⟩ Look for the monsters!

pages 30 and 31

Can you see Bobo? This is his classroom. It is a funny classroom with a lot of monsters!
Where's Jeremy?

Now look for:

a monster with two heads, a monster with three arms, a fat monster, a tall monster, a monster with ten legs, a purple monster, a girl with blue hair, a boy with two pairs of glasses, a boy with red eyes

⟨8⟩ Listen and join!

page 32

Listen to the CD and join the dots.

Join the red numbers: 1, 3, 7, 5, 9, 6, 2, 4, 8, 10.

Join the blue numbers: 9, 7, 3, 2,

4, 10, 1, 5, 6, 8.

Join the green numbers: 5, 9, 4, 7, 6, 3, 1, 2, 8, 10.

Join the black numbers: 2, 4, 6, 8, 10, 1, 3, 5, 7, 9.

Join the purple numbers: 3, 5, 2, 4, 1, 6.

Cut-out page

1. Cut the 4 strips along the lines.
2. Glue B to B and C to C.
3. Glue A to A.
4. Fold the 2 strips together.
5. Glue D to D.
6. Pull the spring.
7. Push the spring back.
8. Cut the box piece along the lines.
9. Fold along the dashed lines.
10. Glue E, F, G, H, I, J and make the box.
11. Cut Jack along the lines.
12. Glue the spring to Jack (D to D) and to the bottom of the box.
13. Close the box.
14. Open the box.

GLOSSARY

age 年齡

antennae 觸角

arm 胳臂

because 因為

brown 棕色

bug 臭蟲

bullies 強橫霸道的人

classroom 教室

crest （鳥、獸的）肉冠，羽冠

dangerous 危險的

dwarf 矮子

eye 眼睛

falls down 摔倒

fat 肥胖的

funny 有趣的

glasses 眼鏡

go away 走開

hair 頭髮

handsome 英俊的

horn （牛、羊、鹿等動物的）角

legs 腿

like 像…一樣

little 矮小的

monster 怪物

mystery 神秘的事物

nice 可愛的

nose 鼻子

orange 橙色

pairs （幾）雙，（幾）對

pink 粉紅色

purple 紫色

real 真正

scared 害怕的

short 矮的

silly 傻的

spider 蜘蛛

spike 尖狀物

tail 尾巴

thin 瘦的

tongue 舌頭

toy 玩具

trunk 象鼻，長鼻

wears 穿，戴

wing 翅膀

yellow 黃色

QUALITY ENGLISH CLUB

Membership Application Form

QUALITY ENGLISH CLUB is for those who love English reading and seek for better English to share and learn with fun together.

Benefits offered: - *Membership Card*   - *English learning activities*
                         - *English learning e-forum*   - *Surprise gift and more...*

Simply fill out the application form below and fax it back to 2565 1113 or send it back to the address at the back.

Join Now! It's FREE exclusively for readers who have purchased books on Quality English Learning published by the Commercial Press**!**

(Please fill out the form with **BLOCK LETTERS**.)

The title of book(s) /book set(s) that you have purchased: _____

English Name: _____ (Surname) _____ (Given Name)

Chinese Name: _____

Address:

Tel: _____   Fax: _____

Email: _____
                                    (Login password for e-forum will be sent to this email address.)

Sex: ❏ Male   ❏ Female

Education Background: ❏ Kindergarten   ❏ Primary 1-3   ❏ Primary 4-6
                     ❏ Junior Secondary Education (F1-3)   ❏ Senior Secondary Education (F4-5)
                     ❏ Matriculation   ❏ College   ❏ University or above

Age: ❏ 3 - 6   ❏ 6 - 9   ❏ 10 - 12   ❏ 13 - 15   ❏ 16 - 18
       ❏ 19 - 24   ❏ 25 - 34   ❏ 35 - 44   ❏ 45 - 54   ❏ 55 or above

Occupation: ❏ Student   ❏ Teacher   ❏ White Collar   ❏ Blue Collar
              ❏ Professional   ❏ Manager   ❏ Business Owner   ❏ Housewife
              ❏ Others (please specify: _____)

As a member, what would you like **QUALITY ENGLISH CLUB** to offer:

❏ Member gathering/ party   ❏ English class with native teacher   ❏ English competition
❏ Newsletter   ❏ Online sharing   ❏ Book fair
❏ Book discount   ❏ Others (please specify: _____)

Other suggestions to **QUALITY ENGLISH CLUB**: _____

_____

Please sign here: _____   (Date: _____)

Visit us at Quality English Learning Online http://publish.commercialpress.com.hk/qel

Stamp
Here

QUALITY ENGLISH CLUB

The Commercial Press (Hong Kong) Ltd.
8/F, Eastern Central Plaza,
3 Yiu Hing Road, Shau Kei Wan,
Hong Kong

THE COMMERCIAL PRESS (H.K.) LTD.